I Hate Trig!

A Practical Guide to Understanding Trigonometry

Jesse Moland, Jr.

I Hate Trig!
by Jesse Moland, Jr.

Right Angle Trigonometry

What exactly is trigonometry? It is a branch of mathematics that deals with the sides and angles of triangles, their relationships and other functions.

We will begin with right angle trigonometry. Once you understand right angle trigonometry, trigonometry in general will be easy.

A right angle triangle is one that has a 90° angle. A unique property of right angle triangles is the fact that they adhere to the Pythagorean theorem.

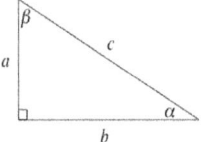

$c^2 = a^2 + b^2$ - the square of the hypotenuse is equal to the sum of the squares of the other two sides.

To help us understand further, we will give the other two sides names: opposite and adjacent. Which is which? Well, that depends on what angle you are talking about.

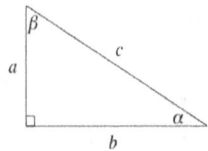

In this triangle, side *a* is opposite of angle α and side *b* is adjacent to angle α. But, if we are working with angle β, side *b* is the opposite side and side *a* is the adjacent side.

Remember, trigonometry deals with the relationship between sides and angles.

Special Triangles

There are two special triangles that you must also commit to memory. Both of these right angle triangles adhere to the Pythagorean theorem:

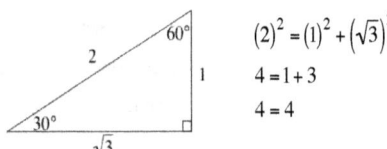

$$(2)^2 = (1)^2 + \left(\sqrt{3}\right)^2$$
$$4 = 1 + 3$$
$$4 = 4$$

$$\left(\sqrt{2}\right)^2 = \left(1\right)^2 + \left(1\right)^2$$
$$2 = 1 + 1$$
$$2 = 2$$

From the two special triangles, we see three distinct angles: 30°, 45°, and 60°.

Basic Trigonometric Functions

Now that you understand the relationship between sides and angles in a right angle triangle, you are ready to learn the trigonometric ratios. The three basic trigonometric ratios or functions are sine (sin), cosine (cos), and tangent (tan). Sine is the ratio of the opposite side over the hypotenuse. Cosine is the ratio of the adjacent side over the hypotenuse. Tangent is the ratio of the opposite side over the adjacent side. Remember, which side is which depends on the angle. We will use Θ (theta) as our variable angle.

$$\sin\Theta = \frac{opposite}{hypotenuse} \qquad \cos\Theta = \frac{adjacent}{hypotenuse} \qquad \tan\Theta = \frac{opposite}{adjacent}$$

3

To fully understand trigonometry, you must understand and memorize these three ratios.

Combining what you have learned about the special triangles and the trigonometric ratios, you should be able to find the sine, cosine, and tangent of 30°, 45°, and 60°.

$$\sin 30° = \frac{opposite}{hypotenuse} = \frac{1}{2} \qquad \sin 45° = \frac{opposite}{hypotenuse} = \frac{1}{\sqrt{2}}$$

$$\cos 30° = \frac{adjacent}{hypotenuse} = \frac{\sqrt{3}}{2} \qquad \cos 45° = \frac{adjacent}{hypotenuse} = \frac{1}{\sqrt{2}}$$

$$\tan 30° = \frac{opposite}{adjacent} = \frac{1}{\sqrt{3}} \qquad \tan 45° = \frac{opposite}{adjacent} = \frac{1}{1}$$

$$\sin 60° = \frac{opposite}{hypotenuse} = \frac{\sqrt{3}}{2}$$

$$\cos 60° = \frac{adjacent}{hypotenuse} = \frac{1}{2}$$

$$\tan 60° = \frac{opposite}{adjacent} = \frac{\sqrt{3}}{1}$$

You will use these so often that you will eventually memorize the values of these ratios. Remember, the process is nothing special. You are simply writing a ratio. Once you know the ratios for sine, cosine, and tangent, you can solve for the ratio of any angle in a right angle triangle.

Reciprocal Trigonometric Functions

In addition to the three basic trigonometric functions, there are three reciprocal functions: cosecant (csc), secant (sec), and cotangent (cot).

$$\csc\Theta = \frac{hypotenuse}{opposite} \qquad\qquad \csc\Theta = \frac{1}{\sin\Theta}$$

$$\sec\Theta = \frac{hypotenuse}{adjacent} \qquad\qquad \sec\Theta = \frac{1}{\cos\Theta}$$

$$\cot\Theta = \frac{adjacent}{opposite} \qquad\qquad \cot\Theta = \frac{1}{\tan\Theta}$$

Notice that the reciprocal is not just that of the sides, but it can also be the reciprocal of the function itself.

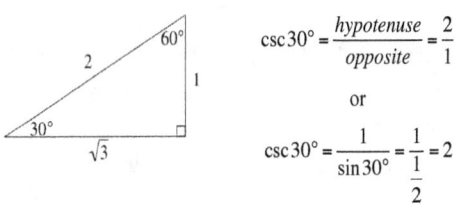

$$\csc 30° = \frac{hypotenuse}{opposite} = \frac{2}{1}$$

or

$$\csc 30° = \frac{1}{\sin 30°} = \frac{1}{\frac{1}{2}} = 2$$

You can now find the sine, cosine, tangent, cosecant, secant, and cotangent of 30°, 45°, and 60° angles.

$$\csc 30° = \frac{hypotenuse}{opposite} = \frac{2}{1} \qquad \csc 45° = \frac{hypotenuse}{opposite} = \frac{\sqrt{2}}{1}$$

$$\sec 30° = \frac{hypotenuse}{adjacent} = \frac{2}{\sqrt{3}} \qquad \sec 45° = \frac{hypotenuse}{adjacent} = \frac{\sqrt{2}}{1}$$

$$\cot 30° = \frac{adjacent}{opposite} = \frac{\sqrt{3}}{1} \qquad \cot 45° = \frac{adjacent}{opposite} = \frac{1}{1}$$

$$\csc 60° = \frac{hypotenuse}{opposite} = \frac{2}{\sqrt{3}}$$

$$\sec 60° = \frac{hypotenuse}{adjacent} = \frac{2}{1}$$

$$\cot 60° = \frac{adjacent}{opposite} = \frac{1}{\sqrt{3}}$$

Related Angles

Now that you can find the six trigonometric ratios of the angles in the special triangles, you can use related angles to solve for even more angles.

Related angles are always acute (less than 90°) and are always made with respect to the x-axis. For example,

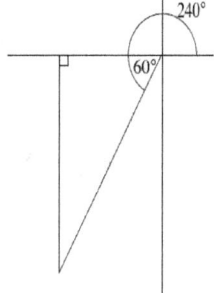

240° is the same thing as 60° in the third
quadrant. 60° is the related angle.

The following figures show how to find the related angle in
each quadrant.

Quadrant I

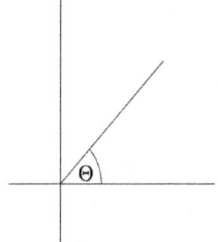

No related angles. All angles are acute.

Quadrant II

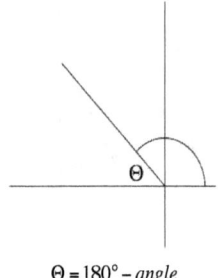

$$\Theta = 180° - angle$$

Quadrant III

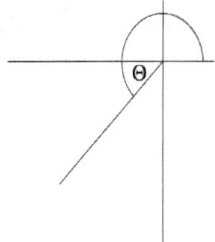

$$\Theta = angle - 180°$$

Quadrant IV

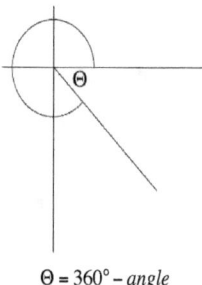

$$\Theta = 360° - angle$$

By placing the two special triangles in different quadrants, you will be able to solve for any angle that is related to 30°, 45°, and 60°. The value of the ratio will be the same regardless of the quadrant. The sign, however, will be different.

Signs of Trigonometric Functions

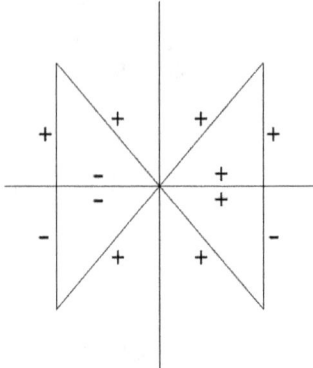

Notice that the hypotenuse is always positive. Let's examine each quadrant more closely and see how to find the related angle and apply the proper sign.

Quadrant I

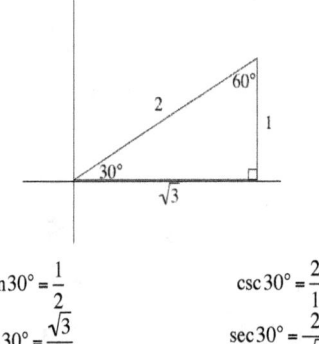

$$\sin 30° = \frac{1}{2}$$

$$\csc 30° = \frac{2}{1}$$

$$\cos 30° = \frac{\sqrt{3}}{2}$$

$$\sec 30° = \frac{2}{\sqrt{3}}$$

$$\tan 30° = \frac{1}{\sqrt{3}}$$

$$\cot 30° = \frac{\sqrt{3}}{1}$$

All trigonometric functions are positive in the first quadrant.

Quadrant II

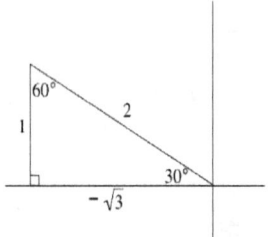

$$\sin 30° = \frac{1}{2}$$

$$\csc 30° = \frac{2}{1}$$

$$\cos 30° = -\frac{\sqrt{3}}{2}$$

$$\sec 30° = -\frac{2}{\sqrt{3}}$$

$$\tan 30° = -\frac{1}{\sqrt{3}}$$

$$\cot 30° = -\frac{\sqrt{3}}{1}$$

Sine and cosecant are positive in the second quadrant, while cosine, secant, tangent, and cotangent are negative.

Quadrant III

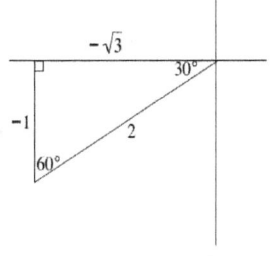

$$\sin 30° = -\frac{1}{2}$$

$$\cos 30° = -\frac{\sqrt{3}}{2}$$

$$\tan 30° = \frac{1}{\sqrt{3}}$$

$$\csc 30° = -\frac{2}{1}$$

$$\sec 30° = -\frac{2}{\sqrt{3}}$$

$$\cot 30° = \frac{\sqrt{3}}{1}$$

Sine, cosecant, cosine, and cosecant are negative in the third quadrant, while tangent and cotangent are positive.

13

Quadrant IV

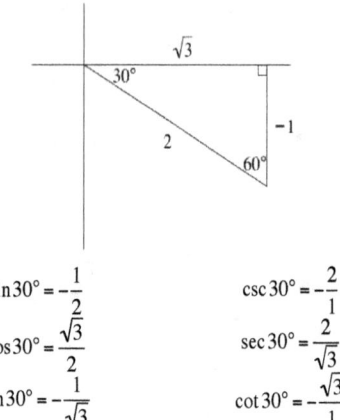

$$\sin 30° = -\frac{1}{2}$$

$$\cos 30° = \frac{\sqrt{3}}{2}$$

$$\tan 30° = -\frac{1}{\sqrt{3}}$$

$$\csc 30° = -\frac{2}{1}$$

$$\sec 30° = \frac{2}{\sqrt{3}}$$

$$\cot 30° = -\frac{\sqrt{3}}{1}$$

Sine, cosecant, tangent, and cotangent are negative in the fourth quadrant, while cosine and secant are positive.

Each basic and reciprocal trigonometric function is positive in two quadrants and negative in two quadrants. (Remember, each reciprocal function has the same sign as its corresponding basic function.)

$\sin\Theta$ is $+$ $\cos\Theta$ is $-$ $\tan\Theta$ is $-$	$\sin\Theta$ is $+$ $\cos\Theta$ is $+$ $\tan\Theta$ is $+$
$\sin\Theta$ is $-$ $\cos\Theta$ is $-$ $\tan\Theta$ is $+$	$\sin\Theta$ is $-$ $\cos\Theta$ is $+$ $\tan\Theta$ is $-$

Quadrantal Angles

The angles that define the quadrants, 0°, 90°, 180°, 270°, and 360°, also have trigonometric ratios. While the ratios for the other angles we have seen can be found using simple triangles, the values of the ratios for the quadrantal angles, once developed, should be memorized.

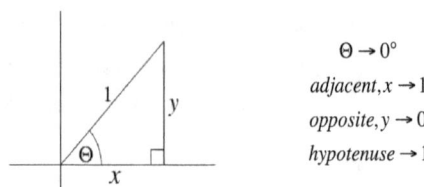

$$\Theta \to 0°$$
$$adjacent, x \to 1$$
$$opposite, y \to 0$$
$$hypotenuse \to 1$$

As Θ approaches 0°, the length of the opposite side approaches 0. At the same time, the length of the adjacent side approaches the value of the hypotenuse. The length of the hypotenuse remains unchanged.

$$\sin\Theta = \frac{opposite}{hypotenuse} \qquad \qquad \sin0° = \frac{0}{1} = 0$$

$$\cos\Theta = \frac{adjacent}{hypotenuse} \qquad \qquad \cos0° = \frac{1}{1} = 1$$

$$\tan\Theta = \frac{opposite}{adjacent} \qquad \qquad \tan0° = \frac{0}{1} = 0$$

$$\csc\Theta = \frac{hypotenuse}{opposite} \qquad \qquad \csc0° = \frac{1}{0} = DNE$$

$$\sec\Theta = \frac{hypotenuse}{adjacent} \qquad \qquad \sec0° = \frac{1}{1} = 1$$

$$\cot\Theta = \frac{adjacent}{opposite} \qquad \qquad \cot0° = \frac{1}{0} = DNE$$

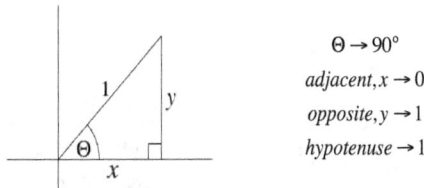

$$\Theta \to 90°$$
$$adjacent, x \to 0$$
$$opposite, y \to 1$$
$$hypotenuse \to 1$$

As Θ approaches 90°, the length of the opposite side approaches the value of the hypotenuse. At the same time, the length of the adjacent side approaches 0. The length of the hypotenuse remains unchanged.

$$\sin\Theta = \frac{opposite}{hypotenuse} \qquad\qquad \sin 90° = \frac{1}{1} = 1$$

$$\cos\Theta = \frac{adjacent}{hypotenuse} \qquad\qquad \cos 90° = \frac{0}{1} = 0$$

$$\tan\Theta = \frac{opposite}{adjacent} \qquad\qquad \tan 90° = \frac{1}{0} = DNE$$

$$\csc\Theta = \frac{hypotenuse}{opposite} \qquad\qquad \csc 90° = \frac{1}{1} = 1$$

$$\sec\Theta = \frac{hypotenuse}{adjacent} \qquad\qquad \sec 90° = \frac{1}{0} = DNE$$

$$\cot\Theta = \frac{adjacent}{opposite} \qquad\qquad \cot 90° = \frac{0}{1} = 0$$

17

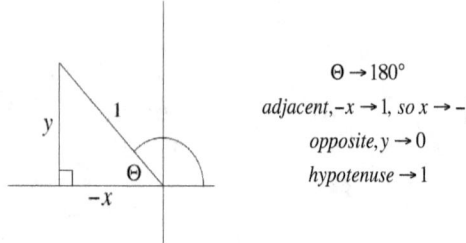

$$\Theta \to 180°$$
$$adjacent, -x \to 1, \ so \ x \to -1$$
$$opposite, y \to 0$$
$$hypotenuse \to 1$$

As Θ approaches 180°, the length of the opposite side approaches 0. At the same time, the length of the adjacent side approaches the value of the hypotenuse, but because of the quadrant, it is negative. The length of the hypotenuse remains unchanged.

$$\sin\Theta = \frac{opposite}{hypotenuse} \qquad\qquad \sin 180° = \frac{0}{1} = 1$$

$$\cos\Theta = \frac{adjacent}{hypotenuse} \qquad\qquad \cos 180° = \frac{-1}{1} = -1$$

$$\tan\Theta = \frac{opposite}{adjacent} \qquad\qquad \tan 180° = \frac{0}{-1} = 0$$

$$\csc\Theta = \frac{hypotenuse}{opposite} \qquad\qquad \csc 180° = \frac{1}{0} = DNE$$

$$\sec\Theta = \frac{hypotenuse}{adjacent} \qquad\qquad \sec 180° = \frac{1}{-1} = -1$$

$$\cot\Theta = \frac{adjacent}{opposite} \qquad\qquad \cot 180° = \frac{-1}{0} = DNE$$

18

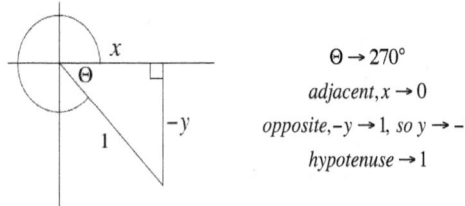

$$\Theta \to 270°$$
$$adjacent, x \to 0$$
$$opposite, -y \to 1, \; so \; y \to -1$$
$$hypotenuse \to 1$$

As Θ approaches 270°, the length of the opposite side approaches the value of the hypotenuse, but because of the quadrant, it is negative. At the same time, the length of the adjacent side approaches 0. The length of the hypotenuse remains unchanged.

$$\sin\Theta = \frac{opposite}{hypotenuse} \qquad\qquad \sin 270° = \frac{-1}{1} = -1$$

$$\cos\Theta = \frac{adjacent}{hypotenuse} \qquad\qquad \cos 270° = \frac{0}{1} = 0$$

$$\tan\Theta = \frac{opposite}{adjacent} \qquad\qquad \tan 270° = \frac{-1}{0} = DNE$$

$$\csc\Theta = \frac{hypotenuse}{opposite} \qquad\qquad \csc 270° = \frac{1}{-1} = -1$$

$$\sec\Theta = \frac{hypotenuse}{adjacent} \qquad\qquad \sec 270° = \frac{1}{0} = DNE$$

$$\cot\Theta = \frac{adjacent}{opposite} \qquad\qquad \cot 270° = \frac{0}{-1} = 0$$

There is no need for us to solve using an angle of 360° because it has the same values as 0°. This is because 0° and 360° are coterminal angles.

Coterminal Angles (Angles Greater Than 360°)

All angles have an initial side and a terminal side. If two angles share the same terminal side, they are said to be coterminal.

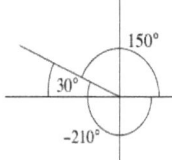

150° and -210° are coterminal angles. Notice that the related angle for each is 30° in the second quadrant.

Angles that are greater than 360° or less than -360° should be written in terms of a coterminal angle. This can be accomplished quickly by either adding or subtracting a multiple of 360°. Once this coterminal angle is found, the related angle can be found.

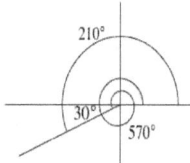

Since 570° is greater than 360°, we will subtract 360° to find the coterminal angle.

$$570° - 360° = 210°$$

Now we can find the related angle.

$$210° - 180° = 30°$$

So 570° is the same as 30° in the third quadrant.

Inverse Trigonometric Functions

Remember that the six basic trigonometric functions give you a ratio based on a given angle. How do you find the angle given the ratio?

The inverse trigonometric functions are Arcsin, Arccos, Arctan, Arccsc, Arcsec, and Arccot. Notice that they all have the prefix "Arc-". This basically means "the angle whose". Therefore Arcsin means "the angle whose sine is".

Each inverse trigonometric function is defined as positive in one quadrant and negative in one quadrant.

Arcsin and Arccsc are positive in the first quadrant and negative in the fourth quadrant.

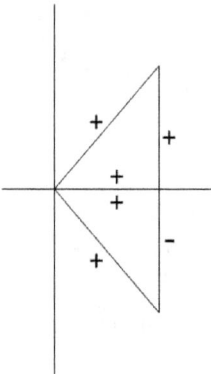

Arccos and Arcsec are positive in the first quadrant and negative in the second quadrant.

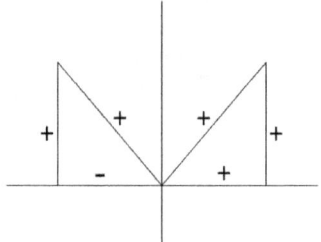

Arctan and Arccot are positive in the first quadrant and negative in the fourth quadrant.

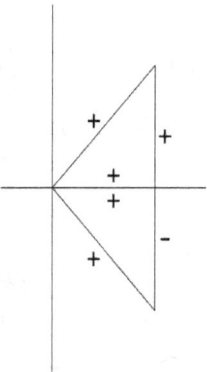

Using this information, simply place the ratio in the correct quadrant and solve for the angle. Let's solve $\text{Arcsin}\left(-\dfrac{1}{2}\right)$.

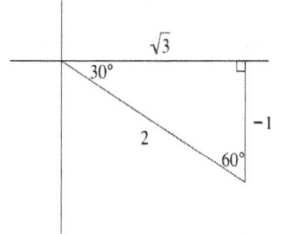

Since Arcsin is negative in the fourth quadrant, we draw a triangle there and label the opposite side and hypotenuse.

23

Once the sides are labeled, we see that this is one of our special triangles, and the angle is -30°. Therefore, $\text{Arcsin}\left(-\dfrac{1}{2}\right) = -30°$ or $330°$.

Inverse trigonometric functions are easy to use when the ratio clearly comes from a special triangle. In some instances, it may be necessary to rationalize the numerator. For example, $-\dfrac{\sqrt{3}}{3}$ should remind you of a couple of trigonometric ratios you have worked with, but not exactly. If we rationalize the numerator, $-\dfrac{\sqrt{3}}{3}\left(\dfrac{\sqrt{3}}{\sqrt{3}}\right) = -\dfrac{3}{3\sqrt{3}} = -\dfrac{1}{\sqrt{3}}$, we see a ratio that is clearly $\tan\left(-30°\right)$ or $\tan\left(330°\right)$.

Powers of Trigonometric Functions

Just as numbers and variables can be squared, functions can be squared as well. There are three notations that can be used to represent this:

$$\sin\Theta^2 \qquad\qquad \left(\sin\Theta\right)^2 \qquad\qquad \sin^2\Theta$$

The third notation is the most common, but the second notation more completely describes what is going on – squaring the value of $\sin\Theta$.

$$\sin^2 60° =$$
$$\left(\sin 60°\right)^2 =$$
$$\left(\frac{\sqrt{3}}{2}\right)^2 =$$
$$\frac{3}{4}$$

Once you find $\sin 60°$, simply square that value to find your final answer.

Trigonometric Equations

As with all other functions, trigonometric functions can be used in an equation. The unknown variable is the argument of the trigonometric function which you must isolate in order to solve.

Let's solve $\cos\Theta - \dfrac{\sqrt{3}}{2} = 0$, $0° \le \Theta < 360°$.

Rearrange the equation to isolate our trigonometric function. Label the sides of the triangles appropriately to get the answers.

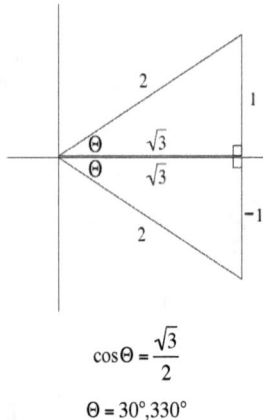

$$\cos\Theta = \frac{\sqrt{3}}{2}$$

$$\Theta = 30°,330°$$

We end up with two answers because cosine is positive in two quadrants – first and fourth.

Don't be distracted by the condition that appears after the problem. It simply tells you that all of your answers for Θ must be greater than or equal to 0° and less than 360°.

If your trigonometric equation includes squared functions, treat it as you would any quadratic equation. Watch the similarities between these two equations.

$$x^2 - x = 0$$
$$x(x-1) = 0$$
$$x = 0 \ \& \ x - 1 = 0$$
$$x = 0,1$$
$$\sin^2 x - \sin x = 0$$
$$\sin x(\sin x - 1) = 0$$
$$\sin x = 0 \ \& \ \sin x - 1 = 0$$
$$x = 0°,180° \ \& \ x = 90°$$

Trigonometric Identities

The following is a list of all of the trigonometric identities that must be memorized in order to effectively solve this type of problem.

Reciprocal Functions

$$\sin\Theta = \frac{1}{\csc\Theta} \qquad \cos\Theta = \frac{1}{\sec\Theta} \qquad \tan\Theta = \frac{1}{\cot\Theta}$$

$$\csc\Theta = \frac{1}{\sin\Theta} \qquad \sec\Theta = \frac{1}{\cos\Theta} \qquad \cot\Theta = \frac{1}{\tan\Theta}$$

Functions of Negative Angles

$$\sin(-\Theta) = -\sin\Theta \qquad \cos(-\Theta) = \cos\Theta \qquad \tan(-\Theta) = -\tan\Theta$$

$$\csc(-\Theta) = -\csc\Theta \qquad \sec(-\Theta) = \sec\Theta \qquad \cot(-\Theta) = -\cot\Theta$$

General Functions

$$\tan\Theta = \frac{\sin\Theta}{\cos\Theta} \qquad\qquad \cot\Theta = \frac{\cos\Theta}{\sin\Theta}$$

Pythagorean Identities

$$\sin^2\Theta + \cos^2\Theta = 1 \qquad 1 + \cot^2\Theta = \csc^2\Theta \qquad \tan^2\Theta + 1 = \sec^2\Theta$$

$$\sin^2\Theta = 1 - \cos^2\Theta \qquad \cot^2\Theta = \csc^2\Theta - 1 \qquad \tan^2\Theta = \sec^2\Theta - 1$$

$$\cos^2\Theta = 1 - \sin^2\Theta \qquad 1 = \csc^2\Theta - \cot^2\Theta \qquad 1 = \sec^2\Theta - \tan^2\Theta$$

*Notice

$$\frac{\sin^2\Theta + \cos^2\Theta = 1}{\sin^2\Theta} = 1 + \cot^2\Theta = \csc^2\Theta$$

$$\frac{\sin^2\Theta + \cos^2\Theta = 1}{\cos^2\Theta} = \tan^2\Theta + 1 = \sec^2\Theta$$

Sum and Difference Identities

$$\sin(A + B) = \sin A \cos B + \cos A \sin B$$

$$\sin(A - B) = \sin A \cos B - \cos A \sin B$$

$$\cos(A + B) = \cos A \cos B - \sin A \sin B$$

$$\cos(A - B) = \cos A \cos B + \sin A \sin B$$

$$\tan(A + B) = \frac{\tan A + \tan B}{1 - \tan A \tan B}$$

$$\tan(A - B) = \frac{\tan A - \tan B}{1 + \tan A \tan B}$$

Double-angle Identities

$$\sin 2A = 2\sin A \cos A$$

$$\cos 2A = \cos^2 A - \sin^2 A \quad \cos 2A = 1 - 2\sin^2 A \quad \cos 2A = 2\cos^2 A - 1$$

$$\tan 2A = \frac{2\tan A}{1 - \tan^2 A}$$

Half-angle Identities

$$\sin \frac{x}{2} = \pm \sqrt{\frac{1 - \cos x}{2}} \qquad\qquad \cos \frac{x}{2} = \pm \sqrt{\frac{1 + \cos x}{2}}$$

Conclusion

There is no way that you can be successful in trigonometry without memorizing the appropriate ratios, relationships, and identities. They are foundational to understanding and are completely inescapable.

www.ingramcontent.com/pod-product-compliance
Lightning Source LLC
Chambersburg PA
CBHW072046190526
45165CB00018B/1879